Flight Plan to the Flight Deck

Strategies for a Pilot Career

Judy A. Tarver

Published by Cage Consulting, Inc.
Englewood, Colorado

Flight Plan to the Flight Deck: Strategies for a Pilot Career
Judy A. Tarver

Cover Design by Alexander Cannon
Editing and Layout by Pam Ryan

Updated and Revised 1999
Copyright 1997 by Judy A. Tarver
First Printing 1997

Printed in the United States of America
Published by CAGE CONSULTING, INC.

Library of Congress Catalog Card Number: 96-072541
ISBN 0-9642839-2-1

Disclaimer: This book is sold with the understanding that the publisher and the author are not engaged in rendering legal or medical services. If legal or medical expert assistance is required, the services of a competent professional should be sought.

This book is a general information book on how to plan a professional pilot career path. It is understood that the information contained in this book does not guarantee success. The author and publisher shall have neither liability nor responsibility to any person or entity with respect to any loss or damage caused, or alleged to be caused directly or indirectly by the information contained in this book.

If you do not wish to be bound by the above, you may return this book to the publisher for a full refund.

Table of Contents

Aviation publications and seminars.
Your friends.
A word of warning.

Choosing an employer.
Research.
The resume.
Sample resume.
The application.
The cover letter.
Updates.

How hiring criteria are established.
Selection process.
Final selection.

Preparation.
Appearance.
Behavior.
Follow up.
Rejection.

Networking
Services
Associations
Publications

What Every Pilot Wonders:

FOREWORD

When you were a youngster you looked up in the sky and watched the airplanes soar over your home. You sat at the end of the runway at the local airport and dreamed of the day you would fly as a professional pilot. Now you are ready to turn that dream into reality.

Do you know what it takes to be a professional pilot? Do you know how much it costs to become a pilot? Are you ready to invest the money and time it will take to reach that final goal? This book will help you understand those costs of money and time. It will also point out the sometimes hidden pitfalls, as well as the many rewards, of a career as a professional pilot.

This book will be helpful to those of you just starting on your career path. It will also be helpful to those who have completed their education and basic flight training and are ready to begin the professional pilot job search process. Even the seasoned pilot can benefit from many of the helpful hints relating to preparing for the important interview process.

You must learn how to approach your career logically, responsibly, and realistically.

It takes five to ten years of hard work, sacrifice and determination to secure a satisfactory career in the flying business. You will have to take low paying jobs, go through extensive and expensive flight training, and move where the best career opportunities avail themselves. This book won't teach you how to fly. The objective is to help you reach your goal as a professional pilot while avoiding the pitfalls too many well-intentioned aviators have experienced.

I spent more than fifteen years working with major commercial airlines and other aviation-related companies evaluating pilot credentials. I was responsible for reviewing tens of thousands of applications and played a major role in the decision to hire over 7,100 pilots. I saw dreams come true. But I also saw dreams go down the drain because of mistakes that could have been avoided. I saw men and women take extra years to reach their goal because of poor planning or bad advice.

Over the years, through my work with aviation education and other youth aviation organizations, I counseled hundreds of aspiring pilots. What a thrill it is to see them sitting at the flight controls of a major airline or flying corporate executives all over the world. Those industrious young aviators did their research, followed my advice and took years off the time it normally takes to achieve their dream. I also like to think they are safer pilots because they went by the rules.

My goal in writing this book is to help build a foundation for a generation of pilots. Pilots who will be responsible, safe, honest, and dedicated to maintaining the proud captain's image that the flying public deserves and respects.

Let's get started on your Flight Plan to the Flight Deck!

Judy Tarver
Sterling, Virginia

1

ARE YOUR GOALS ACHIEVABLE?

B efore you go any further, ask yourself this important two-part question: "Do I want to be a pilot, or do I have my sights set on being a pilot for a specific company?"

If you want to be a pilot for a specific company, I suggest you look for another career. Remember the old (but true) saying, "Don't put all your eggs in one basket." Too many people set their sights on one company, only to be rejected in the interview process or never get an interview at all. But, if you want to be a professional pilot, you have a fighting chance if you can perform the essential functions of the job, meet the Federal Aviation Administration (FAA) physical and certificate requirements, and keep yourself open to all pilot employment opportunities.

You must be aware, however, of one fact that never changes. The airline industry is cyclical and timing can be everything. Many pilots missed a golden opportunity because the industry was in a slump. Not only were jobs nonexistent, but pilots were being laid off.

Corporate jobs aren't any more stable. When a recession hits, the first asset to go is the aircraft.

Can you afford the risks? Check your pocketbook. Have a backup plan.

You may be reading this book during a downturn in the industry. Airlines are going bankrupt or shrinking. Why would an intelligent, responsible professional want to invest thousands of dollars, and hundreds of hours preparing for a career that may never materialize? Because you love to fly!

You must be realistic. Aviation is a highly competitive career field. You may never work for a major commercial airline or you may never fly executives around the world for that Fortune 500 corporation. You may meet the physical requirements for one company but not another. You may not be able to, or want to, relocate or be away from home for long periods at a time. Maybe you have other interests and only want to fly part-time.

It is vital that you are aware of the realities prior to starting on your career path.

For a moment, put aside the actual flight training you must undertake and think about the following prerequisites for the job of a professional pilot.

CAN YOU MEET THE MEDICAL CERTIFICATION STANDARDS?

To fly as a captain or pilot-in-command (PIC), a pilot must be able to secure a First Class Medical

Certificate. The Federal Aviation Administration (FAA) has specific medical standards that must be met before you can operate an aircraft. Before you even begin flying, go to a reputable Aviation Medical Examiner (AME) and make sure you can meet the FAA requirements for a First Class Medical Certificate.

Here are a few basic things to think about:

A First Class Medical Certificate is good for six months. Companies invest a lot of money in training you. They want to know that you will be around longer than six months and that you are not a safety risk. Many companies apply more stringent criteria than the basic First Class Medical Certificate during the initial employment physical. Therefore, it is in your best interest to heed the following advice.

I recommend you go one step further than completing a First Class Medical—get the most thorough physical you can afford. Granted, you don't need a First Class Medical to be a private pilot, but you will need one when you become a captain or pilot-in-command. If you don't meet the standards now, you will want to reconsider your career path.

There are many aspects of your physical health that are controllable:

Your general health. If you lose your certificate, you lose your job. Then what will you do?

Stay fit. Lose weight. Eat properly. Exercise regularly. Do it now. Start a good fitness program. That's one less worry you will have when you come up against the superhuman candidates competing with you for that precious interview. There's an added benefit—maybe you'll get to fly until retirement and make those big bucks everyone talks about.

Some of the things you want to be conscious of are:

Your cholesterol (have you had it checked lately?), your blood pressure, your weight (is it in proportion to your height?), and your hearing.

Your hearing. Now is the time to turn down that stereo. Airplanes are noisy enough. You don't need to reduce your hearing further for a few minutes of esthetic pleasure. Always wear earplugs around excessive noise. Start doing it today. You can be rejected because you don't meet certain hearing requirements. Some simple examples: you must be able to hear a normal conversation from six feet away with your back to the doctor; you will have to hear tones at different frequencies like 500hz, 2,000hz, and 3,000hz. Your AME can go into more detail about preventing hearing loss.

This may all appear to be common sense. But, when pilots ignore taking the most basic of precautions to protect their health the consequences can be devastating: the loss of a job offer, or even the loss of their pilot license.

There are some things you can't control:

Your height. You must be able to safely reach and operate all essential controls on the aircraft. Most companies do not have published height requirements.

Since people are built differently, measuring your height cannot accurately predict your ability to reach. I was once part of a study to determine how the height criterion should be established at an airline. I am only 5 feet 4 inches tall. I could reach the overhead controls but not the rudders.

Another participant was the same height and could reach the rudders but not the overheads. One other participant, also 5 feet 4 inches, could reach both. In this case, if the airline set a height criterion of 5 feet 6 inches, one individual would lose a job opportunity even though the candidate demonstrated the ability to reach all controls.

A company could use one of several methods to evaluate reach. They might put you in a flight simulator or cockpit of an aircraft so you can demonstrate your ability to reach everything necessary to complete the job responsibilities. Or they may measure your sitting height.

Your vision. One of the basic vision standards set by the FAA for a First Class Medical requires a pilot to have vision correctable to 20/20. If you don't meet the standard requirements please be cautious about using certain techniques which require incisions in the eye, such as radial keratotomy, until you are certain that enough companies accept these types of corrections. (If one company accepts a certain type of correction, but no other company does, you could be greatly limiting your opportunities for employment.)

If you don't meet the vision standards, there is something known as a statement of demonstrated ability (SODA). For example, a pilot may have a color vision deficiency. Under these circumstances there is a separate test that is conducted to insure the pilot can read all the instruments. If passed, the pilot is issued a waiver and allowed to obtain the First Class Medical Certificate.

Your family tree. You can't change your genes but you can take preventive care. If you know that your family has a long history of poor health, you must be extra cautious and go the extra mile to stay in shape.

In July, 1992, the government enacted the Americans With Disabilities Act (ADA) legislation. This law precludes a company from conducting a pre-employment physical prior to extending a job offer. This is intended to keep employers from discriminating against individuals with disabilities.

In spite of this legislation, you still must be able to perform the essential functions of the job and meet the specific requirements of a company and the FAA First Class Medical requirements. For example, one condition of employment may be that you can climb down a rope or be able to jump over a seat. Many airlines publish detailed documentation which outlines their requirements. No company will hire an individual who has a health or physical problem that will compromise safety.

CAN YOU AFFORD TO BE A PILOT?

By the time you land that job with a major airline or corporation you can expect to spend from $60,000 to $100,000 (or more) for your education and all the certificates and ratings required to make you a competitive candidate.

Not only is training expensive, but guess what—you probably won't be seeing any high dollar jobs for a long time. Most charter and commuter companies pay very little compared to the large commercial airlines or stable corporations. That makes paying off your student and flight training loans even more difficult.

Some examples of average annual salaries for flying jobs (upper salaries are normally captain positions):

✈ Small charter or small corporate $16,000 – $53,000
✈ Regional Airlines/Commuter $19,000 – $70,000
✈ National Airlines $20,000 – $112,000
✈ Corporate Jets $18,000 – $150,000
✈ Major Airlines $24,000 – $130,000
✈ Global Airlines $24,000 – $200,000

To save money many commuter, regional, and national carriers are adopting the practice of hiring only pilots who are prescreened and trained by an external training center. In the industry this type of practice is known as "pay for training." A few carriers will reimburse you for the training costs over a period of years if you stay with the company, however, that is not the standard.

A "pay for training" contractor will prescreen you by putting you through a selection process to make sure that you meet the minimum criteria to be successful in the training program. The average charge for the prescreening runs about $300. This is typically nonrefundable and there is no guarantee you will be selected to attend the school. After that, you could pay up to $10,000 for the training on a specific aircraft. Whether the "pay for training" approach will continue is unknown. A great deal will depend on the future availability of pilots. When the pilot pool is shallow, companies must spend more money to entice qualified applicants—this could include paying for the pilot's training.

DO YOU HAVE SEVERAL YEARS TO PREPARE?

By now you realize that if you want to be a professional pilot you will have to get serious about your career and concentrate on your health and strategic career planning. It is time for you to buckle down and work long hours, realize that you may be away from home more than you want, and that you may have little control over your schedule, or where you are going to live.

If you start flying while you are in college, you may be able to get a job with a major airline in five to ten years. Because of the cyclical nature of the industry, when hiring was booming, as it was in the mid 1980s, a good pilot could expect to land a job with a major company with little worry. However, as hiring slowed down in the early 1990s, jobs were almost nonexistent and even "top gun" superstars couldn't get an airline job. The late 1990s into the year 2000 show promise for aspiring pilots.

WHAT IS YOUR TIMELINE?

Now is also the time to prepare a timeline for achieving your professional pilot qualifications.

First, let's look at the basic certificate requirements that you must receive from the FAA in order to get a job as a professional pilot. You must pass a written examination administered by the FAA and demonstrate certain skills to obtain each certificate:

STUDENT PILOT. You must be at least 16 years old and hold at least a current Third Class Medical Certificate. You must obtain this certificate in order to fly solo. You must receive instruction in basic flight techniques in the aircraft to be flown on your solo flight. Some of the areas included: flight preparation procedures, takeoff and landings, straight and level flight, climbs and climbing turns. After showing competency in the required areas you may then be recommended for your private pilot certificate. You may not carry passengers or get paid as a pilot.

PRIVATE PILOT. You must be at least 17 years old, and hold at least a current Third Class Medical Certificate. In most cases you must have at least 40 hours of approved flight instruction and solo flight time. You must show proficiency in preflight operations, traffic pattern operation, flight at slow airspeeds, crosswind takeoffs and landings, night flying, etc. You may not get paid for being a pilot.

COMMERCIAL PILOT. You must be at least 18 years of age and hold at least a current Second Class Medical Certificate. You

Plan your training timeline carefully.

must have at least 250 hours of flight time, which includes 100 hours of pilot-in-command time. You must also hold an instrument rating. You may get paid for being a pilot.

AIR TRANSPORT PILOT (ATP). You must be at least 23 years old and hold a current First Class Medical Certificate. You must have at least 1,500 hours of flight time that includes at least 500 hours of cross-country flight time, 100 of night flight time, and 75 hours of instrument flight time.

You should expect to receive your student and private pilot's certificates during your first year in college. Over the course of the next three years you will take courses that will lead to your commercial certificate with a multi-engine rating and your certified flight instructor certificate. Many students then work as flight instructors during their last years in college.

Remember, this is only a sample timeline. Your progress may vary.

Here is an example of a timeline for a pilot who attends an aviation college:

	College	Year after Graduation	Beginning 2nd Year after College Graduation	
Total Time	0–400	400–1200	1200–1500	1500+
Certificates	Student, Private, Instrument, Commercial, Multi-engine, CFI		Take the ATP written	ATP Mel/Sel
Multi-engine	0	200	350	350+
Job possibilities	Part-time flight Instructor	Full-time flight instructor		
Job search		Send out resumes to commuter/regional airlines		

As you can see, it may be slow going at the beginning of your training. You will probably be going to school or working at another job to pay for your training. The first jump in hours will most likely be when you begin flight instructing. The second jump in hours will be much larger when you get your first pilot job (charter pilot, night cargo, etc.).

STARTING LATE.

Pilots over the age of 35 frequently ask me if they are too old to start this career. Age is considered a factor in selection only as it relates to experience. It does not matter when you started your career. What matters is what you did once you decided to take that path. If you are thinking about beginning a pilot career later in life, you will have additional hurdles to overcome. It is important to consider these points:

- ✈ You may not get a good return on your training investment.
- ✈ You may never reach the position of captain.
- ✈ You may never get a job with a major or regional airline; and if you do, you may never become vested in their retirement program.
- ✈ Your earning potential will be lower because the time and experience building companies (commuter, small cargo) typically pay minimum salaries.

The pilots who succeeded in this career later in life pursued their flight time and ratings vigorously once they made the decision.

Commercial aviation is a great opportunity, however, for retired (typically older) military pilots who are looking for a second career.

Don't let the glamorous image we all have of pilots cloud your vision. Take this opportunity to evaluate the major obstacles up front so you won't begin this career with unrealistic expectations.

Being a pilot is the greatest job in the world. Pilots are unique in that the desire to soar is overpowering. If you persevere, the end result will make the hard work and sacrifice well worth the effort.

Notes

2

SENIORITY and THE AIRLINES

SENIORITY.

You will hear the word "seniority" used constantly throughout your flying career. Within the airline industry, with few exceptions, the whole promotional, compensation and benefit system is based upon seniority. It is a way of life and where you fall in the seniority system will have a major impact on the economic and promotional progression of your career.

What is seniority? *A position of precedence over others by reason of a longer span of service.*

In the airline business, positions are assigned to pilots in the order of the date they are hired or, in some cases, the date they successfully complete initial training. When new positions become available they are assigned by seniority. In other words, the next person in seniority above you would get the first opportunity to go the next level, perhaps to captain or to a larger aircraft. While you, simply because you were hired a week after this person, must stay in your current position (perhaps co-pilot or flight engineer).

Where you live and your flying schedules are all based upon your seniority. Perhaps you are hired by an airline that has a base in Chicago, your hometown. You want to be based in Chicago. However, because you do not have enough seniority you may have to fly your trips out of the company's less senior base in New York.

Let me tell you why understanding the seniority system is so important and why you do not want to defer a class date with a company where you hope to spend the better part of your professional life. When the industry is in a slump and pilots are not upgrading or moving to other equipment, they may spend many years flying in the same position, or on the same aircraft. If there is a furlough (the same as being laid off), the last pilots hired are the first to be laid off. I know people who were laid off for three or four years. In the 1970s some pilots were furloughed for up to eight years.

One number in the seniority system can be the difference between upgrading to captain quickly or being furloughed. It can have a large economical impact on your future earning potential.

There are advantages to the seniority system.

+ You always know where you stand when it comes to upgrading your position.
+ As long as you pass your company checkrides, you will upgrade when opportunities become available.
+ You have no competition in promotions.

There are disadvantages to the seniority system.

↦ Performance and extra efforts do not help you in promotional opportunities.

↦ Less motivated or less competent pilots will always be ahead of you.

UNIONS AND THE AIRLINES.

Many airline pilots are represented by professional associations or "unions" once they sign on with a particular airline. While it is more typical to see union representation at major or regional airlines, pilots at small companies are sometimes members of unions.

Pilots pay part of their salaries for membership dues. In return the union provides a means for pilots to negotiate compensation and benefit packages as well as satisfactory and safe working conditions.

Unions are also instrumental in working within the industry to provide a voice for the pilots to ensure that air safety is a government and industry priority.

CORPORATE PILOT POSITIONS AND SENIORITY/UNIONS.

Unlike the airlines which employ thousands of pilots, corporate pilots usually work within a much smaller employee group. Under these circumstances you may find many private companies that employ pilots do not use the seniority system. Promotions, upgrades, etc., can be more easily determined on a person-by-person basis when you are dealing with a smaller number of pilots.

Notes

3

ADVANTAGES OF BEING A PILOT

Until now I have been telling you about the problems you might encounter on your way to becoming a professional pilot. Now I am going to tell you some of the advantages.

Since I am not a pilot, I took an unscientific survey to find out why someone would want to pursue this career path. I asked several pilots to tell me why they chose this career. I asked a cross section of the pilot population, from management, to line pilots from major airlines, regional airlines, and corporate flight departments. Not surprisingly, the responses were almost identical. Here is what I learned.

Love of flying. That appears to be inherent in just about every pilot I know. It transcends every other reason.

Pay and Benefits. Or long-term earning potential. Pilots pay their dues when starting out. Once they land a job with a major airline or corporation, the wait is worthwhile.

Live where you want to live. With travel and cockpit jumpseat benefits, you can virtually live anywhere in the world as long as you are on time to fly your trip.

Instant job gratification. Since the majority of pilots are in the career because they love to fly, they are always doing what they love best.

Flexible schedules. With seniority, pilots have the opportunity to select schedules that best fit their personal lifestyle.

Challenging. Flying in and out of major airports in all kinds of conditions tests a pilot each day. Upgrading to new aircraft and different technology also keeps a pilot's senses stimulated.

Everything in your airline career, position, base, aircraft, is based upon seniority.

Freedom. Travel to many different destinations. A pilot is not tied to a desk.

Working with educated and intelligent people. The majority of pilots have a four-year degree or better. Intelligence is highly correlated with success as a pilot.

Fulfillment. It is rewarding to get passengers safely, comfortably, and on-time from point A to B.

Camaraderie. Working with people who love what they do.

Don't take the work home. Except for training, pilots start fresh with each new flight.

Travel benefits. Most passenger airlines provide liberal travel privileges to their employees and their families. They also have reciprocal agreements with other companies.

Time off. Although pilots work some tough hours, they can typically arrange long periods of time off.

Social status. Tell someone you are an airline pilot and watch their interest peak. This career appears to command the respect of many.

Notes

4

DISADVANTAGES OF BEING A PILOT

A long with the good, there can be some disadvantages. Obviously, the good outweighs the bad or there wouldn't be so many pilots out there pursuing the dream.

Time away from home and family. Pilots can be away for long stretches at a time. Some flying can take you away for weeks or even months.

Can't plan in advance for special events. Until you get enough seniority, forget Christmas, Thanksgiving, or special occasions.

Have to pay the price to get there. As mentioned earlier, the cost to achieve the goal is high. Education and training are expensive.

Uniform restrictions. You can't show your creative side. No long hair. You are typically required to wear a hat. You must conform to company standards.

Dependence on health. You must conform to the FAA medical requirements. If you lose your medical, you lose your job.

Job security. In the uncertain aviation industry, companies can go bankrupt due to poor management or furlough during downturns in the economy. Depending upon the company, pilots can retain the right to be recalled up to a certain number of years. When a company declares bankruptcy, which happens frequently in this industry, a pilot has to begin the job search process all over again. When he finds another pilot position, he will start at the bottom of the seniority list with the new company.

Cyclical nature of the industry. Even if a pilot does not get furloughed, when times are bad, movement within the company can be minimal. There are pilots who were flight engineers for over 15 years before they could upgrade to first officer, much less captain.

Schedule changes from month to month. You cannot always control your schedule.

Seniority rules. Under the seniority system, seniority rules, not ability, dedication, or attitude. The only way a pilot might benefit from dedication is to be selected as a check airman or for another management position. Sometimes those positions carry a few perks. The ill-mannered and sometimes less caring pilots still get upgraded before you if they are more senior.

Psychological stress. Mistakes can be deadly. While you may have many hours of pleasurable, nonthreatening flying conditions, when conditions get bad, they can be really bad.

Furloughs. Trying to find work as a furloughed pilot can be difficult. Many companies are hesitant to invest time and money on pilots for fear of losing them when their original company starts recalling. Therefore, in order to secure employment, a pilot may be forced to resign and give up the prized seniority number.

Physical stress. Sometimes, depending on where you fly, the job can be tedious and exhausting. Bad weather, time zone changes, and equipment problems all take a toll on you.

Commuting. Many pilots commute to work. Although commuting is a personal decision and a plus in many cases, you add hours to your day and have to deal with elements like bad weather or full flights, which all too frequently seem to conspire to make your commute difficult.

School doesn't end when you sign on.

Throughout your career you will be going back to the schoolhouse at least every year for training and testing.

Notes

5

YOUR COMPETITIVE ADVANTAGE

When a major airline announces they are hiring, they immediately receive about 3,000 to 5,000 new applications and twice that number of updates.

You must always meet, and often exceed, the minimum requirements before you will get an interview. In order to avoid showing favoritism or discrimination, companies are very careful not to deviate from their basic requirements.

Minimum requirements vary from company to company. Many companies require a certain number of hours and perhaps some experience in a particular aircraft. For example, the typical regional carrier currently likes to see a minimum of 1,200 to 1,500 hours of total time and one major carrier currently requires a type rating in a B-737.

How do you make yourself competitive among the thousands of other pilots seeking this career?

✓ If you meet the requirements and know a pilot from that company, you may have an advantage if he or she will write a recommendation for you. However, the interviewers prefer references from pilots who have flown with you.

✓ Make yourself available in as many ways as possible.

Attend seminars where airline representatives participate. Some of those resources are listed later in Appendix 1, Resources.

Subscribe to services like the Universal Pilot Application Service, Inc. (UPAS) that are used by companies that are hiring.

✓ Always present yourself in the most positive and professional manner.

Use an interview preparation service that will help you enhance your presentation skills and your ability to sell yourself in the most positive way. (Several are listed in Appendix 1, Resources.)

✓ Select your career opportunities carefully. Be sure that you are always moving upward, whether in pay, experience, or aircraft. For example, you want to move from single-engine propeller to multi-engine. Or you might

Take control of your life and health—NOW.

upgrade from piston to turboprop to jet. If you jump from a corporate jet job to a single-engine prop instructor, a red flag goes up for the interviewer, especially when the salary drops significantly.

Do your homework before accepting a position with any company. You want to make sure the company won't be asking you to push the limits of safety.

✓ If it is not too late, keep your grades up in school, especially in flight related classes. If you flunk your flying courses, a red flag goes up quickly. How can you make up for that? Get as much quality experience as possible. Get good references. Ace your FAA written examinations and do well during flight training.

✓ Throughout your career you will need to take many different types of FAA examinations. As I've discussed, you need to pass a test to be a private pilot, a commercial pilot, a flight engineer, etc. When taking FAA written exams, do your very best. A low score indicates a lack of motivation. Many companies take that very seriously.

✓ Take control of your life and your health. Do not put yourself in negative positions that will follow you throughout your career, like getting a DUI or getting fired for cause.

Notes

6

EDUCATION AND TRAINING

In the past, pilots could move through the ranks without a college degree. Today, you are at a disadvantage if you don't have a bachelors degree. Most airlines prefer a four-year degree and can fill their needs using this as part of the minimum criterion.

AVIATION COLLEGE/UNIVERSITY.

If you are just starting out, one option is to go to a college that awards an aviation degree. There are many excellent colleges and universities that offer fine programs. The University Aviation Association (UAA) publishes a catalog of most of the nation's aviation education institutions. You will find information on the UAA in Appendix 1.

Here are some advantages to attending an aviation college:

✓ You can get an excellent education and finish your basic flight training at the same time.

✓ You save time by combining the programs. It can be tough to take a full-time college program and add flight school to it.

✓ Many aviation universities are closely connected with airlines. Representatives from major airlines and industry sit on their advisory councils and provide feedback that is beneficial to the college curriculum.

✓ Many airlines and large corporations have intern programs for college students. Students who demonstrate the motivation and desire to excel in the field go through an intense interview process with the school and with the company for selected positions. They typically get college credit for the work. Once selected, the students typically work for a semester in a related job within the company. This is an excellent way to learn about the company and have them learn about you. Those students who excel are sometimes given preferential treatment in the selection process.

I know pilots who were hired years before the average pilot because they made themselves visible at collegiate aviation functions.

✓ Airlines are looking for well-rounded pilots with good attitudes and team orientation. Crew Resource Management (CRM) is a method of training pilots to effectively communicate in a crew environment. Many colleges are incorporating CRM into the curriculum. This will give you an edge in the job search.

✓Because of aggressive positioning, many colleges do an excellent job of assisting the students in finding flying positions after college. The college placement center or the alumni association are good resources for career opportunities.

As you can see, there are many advantages to securing a degree from an institution with a strong emphasis on aviation. However, there is no requirement to do so. If you have other interests or want a back up career, there is no rule that says you must pursue an aviation degree. You still must complete the flight certification requirements to become a professional pilot.

If you already completed college, or you want a degree that isn't flight related, don't fret. There are many excellent flight schools out there.

FLIGHT SCHOOLS.

Perhaps you already have your degree or you decided that you do not want to attend an aviation college. In this case it will be necessary for you to choose a flight school.

Do your research carefully when searching for a flight school. Try to attend a nationally accredited training institution. These accredited training facilities must meet certain standards recognized by the US Department of Education.

Before you consider a flight school, survey at least three different schools to get a good overview of cost and what is included in the various programs. Some basic questions to ask a potential flight school:

- ✈ How much will the program cost? What is included?
- ✈ What kind of financing or payment programs are available?
- ✈ Is housing provided (if applicable)?
- ✈ How many certificates or ratings does the school provide?
- ✈ How long is the course?
- ✈ How long has the school been in business?
- ✈ What kind of insurance do they have?
- ✈ What is the instructor/student ratio?
- ✈ Will the same instructor be used for the entire training program?
- ✈ What kind of aircraft are used and how many are available for instruction?
- ✈ How do they schedule the flights?
- ✈ Ask for names of previous students and contact them about their experiences.

The National Air Transportation Association and *Flight Training* magazine put out a nice brochure on how to choose a flight school. In fact, *Flight Training* has many other great resources available to help you decide on a training school.

GRADES.

Companies will look at your grades. If they are not good, be prepared to defend the reasons.

Getting good grades in flying related subjects is critical. If you flunked or did poorly in those courses, a company is going to be very leery about hiring you. That's not to say that you won't get hired, but it puts you at a disadvantage and you will have to do some explaining to overcome that obstacle.

Grades in non-flying courses can say a lot about your motivation and dedication. Getting an A in a course you may not be interested in, such as 19th Century English Poets, is a sure indication of academic success and motivation to succeed. So, work hard to succeed in all your courses.

EXTRACURRICULAR ACTIVITIES/ OUTSIDE INTERESTS.

Not only is participation in flying related extracurricular activities fun, but companies look at it as strong motivation towards a flying career.

Belonging to organizations like Alpha Eta Rho (an aviation fraternity), Aviation Exploring (a division of the Boy Scouts of America that provides opportunities for youth between the ages of 14 and 21 to learn more about aviation careers), or the Civil Air Patrol (CAP, also an aviation youth organization), allows you to learn more about aviation in addition to providing networking opportunities. Also, many colleges sponsor flying teams that compete against each other both regionally and nationally. Participating in collegiate competition is an excellent way to enhance your skills, in addition to offering great networking opportunities.

Airlines frequently send representatives to some of the collegiate flying events. I met and hired some great pilots when I attended the National Intercollegiate Flying Association Safety Conference (NIFA SAFECON).

Employers also look for well rounded applicants. Don't disregard other activities like sports, charities, etc.

However, don't go overboard and lose focus on your education. Choose activities that will give you an opportunity to develop your leadership and team building skills.

7

CAREER PATH

There are many ways to gain experience as a pilot. You can reach your goal as a military aviator, as a civilian pilot, or do both.

MILITARY CAREERS.

You may wish to fly your entire career as a pilot in the military or you might take advantage of the excellent training and experience you gain in the military to prepare you for a job in civilian aviation.

Over the years, a large percentage of new-hires at major airlines gained their experience in the military. The military pilot is prescreened medically and psychologically and works in a structured and disciplined environment much like the major airline pilot. These are attributes that make a military pilot desirable.

The obvious plus to the individual pilot is that all this excellent flight training is paid for by the government.

A military career is a good way to gain experience if you keep in mind the following:

✓A four-year degree is required to become a pilot in the military.

✓The military is downsizing. Pilot slots are at a premium.

✓The tour of duty is lengthening. Your commitment to the military can run as long as ten years. Because of the unpredictable nature of the commercial aviation industry, your commitment might be up at a time when the airlines are not hiring. It's a risk.

✓Although there are many flying jobs available for competent pilots, many smaller companies are hesitant to hire pilots with high performance military jet experience for fear of losing them to the major airlines when hiring picks up. That's a valid concern.

For those pilots who want to stay in the military until retirement, there is still an opportunity for a second career as a civilian pilot.

Even though you gain excellent flying experience in the military, keep in mind flying isn't the overriding criterion in pilot selection. Your military career path will be scrutinized along with your flying skills.

+ Were you on track with promotions? If you weren't, be prepared to explain.
+ Did it take you longer than average to complete training?
+ Did you consistently perform unsatisfactorily?

✈ Were your flying hours consistent with those of the rest of your unit, i.e., was the average pilot getting 50 hours a month while you were only getting 20?
✈ Did you have more desk jobs than flying jobs?
✈ Were you ever removed from flying status?

If you are just getting into flying, keep those factors in mind and make a serious effort to avoid those obstacles throughout your career. If you are already completing your military career, and you fall into any of those categories, be prepared to respond to those issues during an interview.

Also remember there is a strong network of military comrades out there. If you did not perform up to standards or were hard to get along with, your military contemporaries will not hesitate to pass that information along to a hiring company. A bad reference is hard to overcome.

On the other hand, the network can sometimes work in your favor. Some airlines look favorably upon recommendations from pilots who know you well and most of all, have flown with you. Information like that gives them a little more feedback that cannot be retrieved from an application.

Airlines also like to see current flying time, so whatever else you do in the military, stay current. When jobs are scarce the pilot who is currently flying has a competitive edge over one who isn't. Securing a guard or reserve slot can help overcome that dilemma.

\REERS.

nay not want a lengthy military career to get
a pilot. Or, for some reason, you may not meet
nents set forth by the military to become a pilot.
)ad. You can still reach the goal, and if things
flow the way they should, perhaps reach it more quickly, as
a civilian pilot.

There were times when the pilot pool was shallow
and airlines hired candidates with only 200 hours of flying
time. If predictions are accurate, that could happen again.
But don't bet your career on it. Plan your career as though
you need thousands of hours of experience. As I mentioned
earlier, when hiring is down, criteria become more
stringent.

I know pilots who were hired by major carriers
when they were 21 years old. Many became captains before
they were 30. These pilots started flight training when they
were 16 years old or even younger. That is not the norm,
however.

A civilian career in aviation takes constant
networking and the ability to see seek out opportunities to
gain flight time and experience. In this chapter I will give
you some courses of action that worked well for many
aspiring pilots.

You might hear people place an emphasis on
"hours." The number of hours you fly is an objective way
to measure experience and is regularly used as a standard
cutoff point when reviewing an application (e.g., minimum
of 1000 hours to be interviewed). There comes a point,
however, when other criteria come into play such as the
quality of your flying, as well as your motivation, attitude,
and the types of jobs you have pursued. The following

information takes you through a sample career progression.[1]

STEP 1: Aviation Education.

Attending an aviation college or university can speed things up. You will probably gain around 250 to 300 hours of flying time while in school while gaining other valuable experience in related fields. That time will increase to around 400 hours if you have a part-time job as a pilot or flight instructor.

STEP 2: Building Hours.

Flight Instructing. The first step toward getting experience. Every pilot must go through training, therefore there are a multitude of openings in the area of flight instructing. Because you can log the hours as an instructor, you get to increase your flying time as you develop excellent skills in training. Many successful civilian pilots gained much of their time by instructing. However, you want to get that experience and move on. Remember that most companies do not typically hire pilots with only single-engine flight instructing time. Get enough time to get you to the next step.

Basic flight instructing is not to be confused with some of the other excellent careers available at flight training centers. These training centers typically hire pilots with specific aircraft experience to conduct training for

---

1   Unfortunately, in this ever-changing industry companies come and go. Therefore, I will not list specific names of companies in my descriptions. However, some organizations that are listed in Appendix 1 keep up-to-date information and I defer to them.

pilots getting ratings in sophisticated multi-engine turboprop or jet aircraft.

Banner towing and crop dusting. Two ways to gain hours. Usually light single-engine experience.

STEP 3: Multi-engine Experience.

Getting multi-engine experience can be a major hurdle. You are at the point at which you realize you have enough experience to know that you made the right decision and are determined that you can do this for a living, but not enough to get a job that can support you. This is the hardest hurdle to overcome. You know you are a good pilot, but how do you convince others? You need multi-engine experience. The following jobs are all opportunities to gain multi-engine experience:

Flight Instructing. Get a multi-engine rating. Multi-engine instructor positions are competitive positions, and are much sought after. However, if you are lucky enough to land one you will gain valuable multi-engine time, plus get paid for instructing.

I know of one creative pilot who purchased a large block of multi-engine time, then collected students on his own and only charged for the time in the aircraft. His initial dollar outlay was substantial, but over time he made all his money back and gained over 150 hours of precious multi-engine time.

Flight Training Companies (Pay for Training). Some carriers select new crewmembers from companies that provide airplane specific training. To participate in this type of program you must first meet the flight hours and flight experience set by the airline. If you meet these requirements, you pay to attend the flight training school. In return you are guaranteed an interview with a specific airline. It is important to be aware that you are not guaranteed a job. If you do not do well in training, or fail your checkride, you will be without a job, and typically with no refund of your training fees.

The requirements for qualifying to participate in these types of schools are usually less stringent than if you were to go directly to an airline. Many aspiring pilots are using this method to get jobs, but it is expensive.

Ferry pilot. Sometimes a business or individual needs to send an airplane to another location and only requires the use of a pilot for that specific trip. Companies will pay pilots on a per trip basis to ferry their aircraft to a specific destination. This is a good way to gain either single-engine or multi-engine time.

Charter pilot. Air charter companies offer a wide range of services. They can carry passengers or freight and they can fall under different FAA regulations than scheduled airlines. These companies fly small to medium sized equipment and provide opportunities for low-time pilots to start off as co-pilots. You can find these companies at most local fixed base operations (FBO). If you get to know people who work at these companies, you may be able to persuade an employer to hire you to fly in a charter operation. You build up experience in single and light twin aircraft. This is not to be confused with the jet charter companies that fly large jet aircraft and have stiff criteria.

STEP 4: Commercial Carrier Experience.

Small commuters. Many small commuters will sometimes hire pilots with low flight hours. It is an excellent way to gain experience and fly in an environment similar to the airlines.

National/regional carriers. In addition to the military, the major airlines look to these carriers for the most competitive candidates. Regional pilots fly in the same environment and, in many cases, under the same federal guidelines as do pilots for the major carriers. For example, they fly in and out of major airports in a high-density environment in all kinds of weather. These carriers also are starting to incorporate more jet aircraft into their fleets. These carriers like to see their applicants have captain time, but don't let that deter you from applying. Airlines do hire co-pilots who haven't upgraded due to seniority constraints. Although their pay and benefits are typically not up to the level of the major/global airlines, it is a good career and often the final goal of many pilots.

Corporate. Some pilots jump from instructing to corporate flying. It is a good way to gain experience, especially if you can get hired by a company that maintains sophisticated jet equipment. Keep in mind that the qualifications to work for a Fortune 500 company or any large corporation can be as stringent as those of the airlines. For example, you may be required to have specific experience as a captain on the type of aircraft the company flies. Corporate pilots do not typically fly as much as airline pilots, usually averaging around 20 to 35 hours a month. These pilots also must be flexible about doing non-flying tasks. A good corporate job can take you to interesting places and introduce you to very exciting people. These companies usually have excellent pay and benefit programs and can be very selective in the hiring

Change jobs only to improve your experience.

process. To learn about the many corporations with flight departments, contact the National Business Aircraft Association (NBAA).

International carriers. These airlines are starting to hire more and more expatriates. Some even have pilots based in the United States. The compensation and benefit packages are typically quite good. The great thing about these carriers is that they usually fly wide body aircraft.

Crew Leasing. Many companies that don't want their own flight department will contract pilots through a crew leasing company. The leasing company actually recruits, trains, and provides the compensation and benefits packages to the pilots. The pilots are typically hired on a contractual basis for six months to several years at a time and are often based outside the United States. Crew leasing companies often do not have a seniority system. Unlike the major airlines, with the right experience you could start off as a captain. The disadvantage of that is if you aren't hired as a captain, upgrading may be difficult.

While progressing through the civilian skies keep these important factors in mind:

✓ Get as much multi-engine experience as possible. Airlines put more emphasis on multi-engine time than single-engine time. Commuters or regional airlines usually want to see at least 250 hours of multi-engine experience.

✓ Pilot-in-command (PIC) or captain time is important. Usually when we talk about pilot-in-command, the larger companies don't look at single-engine propeller PIC time. When the industry is in a slump, multi-engine PIC time can be hard to get because upgrading opportunities become scarce.

✓ Get as many flying certificates as you can as soon as possible after you reach the required hours and experience levels.

✓ Get as much turboprop or jet time as possible.

✓ Get instrument time. Your instrument skills will be assessed in almost every pilot interview.

✓ Flying in and out of major cities in all kinds of weather will enhance your credentials.

✓ Stay current. Companies look closely at your hours in the last six to twelve months prior to your interview.

✓ Be prepared to move. It would be great if you could progress and never have to leave your home, but that is not realistic. Don't limit your opportunities by refusing to accept the possibility of relocating. If you are in a slump, take any reputable flying job you can get until your options improve. I have spoken to people who can't understand why they aren't getting interviews. Many times they are imposing limitations on the jobs they will take and the location of the jobs they will take.

It is not always when you start flying that is important, it is what you do once you decide to pursue it as a career. Although it is an advantage to start as young as possible due to the long-term earning potential, some people don't discover this exciting career until later in life. If you pursue it with the same vigor and plot your path properly, you will have a good chance of fulfilling your goal.

✓ Stay on track. Don't jump around from job to job. Make sure when you change jobs you are always improving your experience. Each job should be a step up to better equipment or more experience. A company is suspicious of an applicant who goes from a good paying job, flying sophisticated equipment to a lower paying one flying a less complex machine. Be prepared to get the third degree if you do that.

Most small companies and commuters realize you are there to gain experience and enhance your credentials for a future airline or major corporate job. In most cases they are very supportive in your efforts, so always be forthright with them and they will provide good references.

✓ Network—make contacts. Everywhere you go, try to meet the people who make hiring decisions at the airports or at fixed base operations (FBO). Although you can find FBOs at major airports, they are typically located at small or ancillary airfields. There you will also find many corporate and small charter and cargo operations.

Stay in touch with as many employees as you can so you find out when people are leaving and when an opening might become available.

Offer to do odd jobs, like cleaning aircraft, to allow pilots and other employees to get to know you. Show them how motivated you are.

Go up to the cockpit when you fly commercially. Talk to the pilots. You never know when someone might be able to give you a lead to a job.

In the early stages of your career send out resumes to every company that owns an aircraft when you meet the minimum qualifications. Eventually someone will give you a break and then you will be on your way to getting more valuable time.

Non-flying jobs.

Should you try to get hired by a carrier in a non-flying position with the hopes of getting on as a pilot? Although this has worked for some employees, I lean more towards saying no. Jobs outside flying detract from the time you need to build hours. It is very difficult to hold down another job while trying to gain flight experience. This could push back the time to get hired, and that will impact your seniority.

It is important to remember that even if an airline offers preferential treatment to employees who want to become pilots, you are only guaranteed an interview. You are not guaranteed a job! You will still be critiqued on flight time and experience. If you are spending all your time in a non-flying related job, and are low on time, the interviewer may wonder how serious you are about a flying career. Even if you are an excellent employee, that is no guarantee you will be hired as a pilot. If you are not currently working for an air carrier, I recommend you find flying jobs instead.

AIR NATIONAL GUARD/RESERVES.

Here's the best-of-both-worlds scenario. Join the Air National Guard or Reserves. Regional and major airlines alike support the notion of hiring Guard/Reserve pilots. In many cases they will hire you with lower time for the following reasons:

✓The military prescreening, training, and experience as mentioned in the section on military careers.

✓You can supplement your civilian flying with military flying and not only stay current, but increase your experience more quickly. This is especially advantageous since many airlines start their new-hires in the flight engineer seat. The airline benefits because you stay current and you benefit because you are increasing your hours, plus you love to fly! You can't do that as a civilian pilot because you are constrained by federal regulations that limit the number of hours you can fly per month and/or year. Military time is excluded from that regulation.

✓Since the airlines are loaded with Guard and Reserve pilots, hiring departments frequently receive feedback from pilots who have flown and worked with prospective applicants. Most airlines place a high value on recommendations from their employees, especially those from pilots who have flown with the applicant.

There are advantages to gaining experience this way, but there are a few things to keep in mind:

✓The Guard and Reserves have a maximum age limit to become a pilot. You must begin pilot training prior to age 26½. Having good contacts in the Guard or Reserve units can be very helpful. Many applicants discovered that, in order to receive the military training they desired, they had to start out working in non-flying positions until they were able to secure a pilot slot.

✓Military recruitment has cycles just like the airline industry. When the military is downsizing or "right sizing," slots can be difficult to secure. You are competing with qualified pilots being released from active duty who want to continue flying. Retiring military pilots, non-flying reservists, or non-flying Guard members are given serious consideration for these coveted positions.

✓ You still have a commitment to the military. Currently, a pilot must serve at least five years after completing pilot training.

✓ You have to spend at least two weeks per year on active duty and, of course, two days per month of your precious days off will be taken up by your military obligations.

✓ You must meet training requirements that could take you away several times per year.

✓ Don't forget the main reason you joined: to serve your country during conflict or national crisis. Many pilots were taken away from their families and jobs during international conflicts such as Desert Storm. Not only did they fly into enemy territory, but many pilots lost a lot of money due to the length of time they were away (they only receive military pay, not their airline salary). Some companies did not provide total pay and benefit packages for the extended time they were away fighting for us.

✓ Some commuters or corporations will hire Guard and Reserve pilots with very low hours. Take advantage of any commuter or corporate opportunity that may arise. Having civilian experience along with the military training will certainly enhance your opportunities to get to a major airline sooner.

8

KEEPING RECORDS

The need to keep good records is imperative in the aviation industry. If you want to get a job, you will need to provide documented information regarding your background throughout your entire career.

LOGBOOKS.

A pilot's most important possession is the logbook. A certificate can be replaced by writing the FAA, but if a logbook is lost, the information can never be fully recovered. The logbook recounts the entire history of a pilot's experience. It can also tell something about the individual's personality and credibility depending on if the logbook is neat and accurate, or filled with corrections and inappropriate comments.

Not only is it required by the FAA that you keep accurate flying records, but as you progress through your career and apply to various companies for new flying opportunities, your logbooks will always be carefully inspected by the prospective employers. The information within, or the lack of it, can have a major impact on the hiring decision.

When completing entries in your logbook, keep the following facts in mind:

✓ Only use logbooks or computer programs that conform with FAA regulations regarding requirements for recording flight time.

✓ BE HONEST. Don't try to beat the system by logging time that could be questioned. It will catch up with you later. If you fly it, log it. If you don't fly it, or if there is any question

Keep personal logbook notes business related, make copies on a regular basis and store them in a safe place.

about the authenticity, don't log it. There is no excuse for recording your hours incorrectly.

Example: As a rule, airlines don't consider simulator time as actual piloting time. Some applicants have logged it as pilot-in-command. The airlines or the FAA might consider that misrepresentation. Always log it where it belongs, in the column for simulator time.

✓ It is very difficult to go back and change a logbook after many years of entries are recorded. If you do have to make corrections, be prepared to explain them.

✓ Enter the information in a concise, legible manner. Make sure all hours are calculated carefully and carried over from page to page correctly. A logbook with numerous errors and miscalculations tells a lot about your attention to detail.

A logbook that has been completely rewritten can be somewhat suspect.

✓Your logbook should match your jobs. If an employer sees an aircraft listed on your resume or application that doesn't coincide with information in your logbook, watch out.

✓Keep your notes clean and business related. I once reviewed a logbook that included comments about the applicant's personal escapades during a very active social life. It made for interesting reading, but was absolutely inappropriate!

✓Always keep an up-to-date copy stored in a safe place.

   If you lose your logbooks, start right away trying to recreate the lost time. Gather pay records from previous employers. Request affidavits from all employers estimating the number of hours flown during your tenure at each company. Get any kind of information you can to verify the hours you flew.

✓Most important, keep your logbooks updated forever. Don't ever think you can stop keeping them because you get a job with a major airline.

Never stop keeping logbooks.

Throughout the years several large major airlines have gone out of business. Many pilots who thought they had a job until retirement lost their jobs. When they applied at other airlines, they had no records of their hours. Recreating their time and getting verification was, to say the least, a challenge.

✓Don't forget to sign each page. A logbook with no signatures is suspicious.

EMPLOYMENT HISTORY.

Employers are very interested in your past work experience. It doesn't matter if you worked in a fast food restaurant, washed aircraft, or were the captain of a B-747. Keep good records. Beginning with your very first job, keep an employment log and an up-to-date resume at all times. It is not unusual to have employers ask you to provide the names and addresses of every employer you have had since you finished high school.

If a company you worked for has gone out of business or if you have any periods of unemployment, be prepared to provide the name and address of someone (other than a relative) who can verify the dates you worked at the bankrupt company, or were unemployed.

You think you'll remember all your jobs. Time has a way of distorting the average person's memory of dates. It is not unusual for a pilot to have had numerous positions over the duration of a flying career. Some jobs may seem insignificant or not related to aviation employment. Keep track of names, addresses, and dates of employment of every job. If a prospective employer finds out that you deliberately omitted any position, that could be considered grounds for rescinding an offer of employment. Even worse you may be terminated for lying on your application.

Just as you conscientiously record the flight time in your logbook, keep accurate logs of your employment. This information will also be incorporated into your final resume (discussed later in the book). Include the following information in your Employment Log (See Appendix 2 for sample Employment Logs):

+ Month, date, and year you started a job.
+ Month, date, and year you left a job.
+ Complete name of the company.

✈ Address and phone number of the company.
✈ Name of your supervisor. Record the names and addresses of coworkers. If the company goes out of business, you may need these contacts to verify your employment.
✈ List your job responsibilities.

EDUCATIONAL INFORMATION.

Employers are also going to ask for information about your education. In some cases, you might even be asked for high school transcripts. Even if you have been out of school for a while and have several employers, you will still be asked for all your college transcripts.

Keep a record of all the colleges and universities you attended. Include dates attended and addresses (See Appendix 3 for a College Log.)

Keep a personal copy of any transcripts with your other important documents. Although some companies will accept a copy of a transcript most companies ask for an official copy of your transcripts. Some companies want the transcripts sent directly to them. Although transfer credits are normally included on subsequent transcripts, that will not be sufficient documentation. Be prepared to contact all schools that you have attended to request official transcripts.

Most institutions charge for official transcripts, so don't forget to find out what the fee is and include it in your request.

A copy of your diploma is nice, but most organizations want actual transcripts in order to review your grades.

Also keep a record of your extracurricular activities. (See Extracurricular Logs in Appendix 4).

JOB SEARCH LOG.

I recommend you keep track of all the companies you contact concerning pilot employment (see Appendix 6 for Job Search Logs). Your Job Search Log should contain the following information:

- ✈ Date of your correspondence to the company.
- ✈ Company name/address/phone number.
- ✈ Contact name/title.
- ✈ What you sent to the company, i.e., original application, any updates, copies of documents.
- ✈ Follow up correspondence or phone calls.
- ✈ Responses. Record any feedback you receive from the company.
- ✈ Interview details.

9

PITFALLS

The rules are simple. The guidelines are logical. Even so, many people don't think the rules apply to them. When it comes to safety and the lives of hundreds of passengers, companies don't take chances. Don't take shortcuts. Although possible, it's very difficult to overcome problems caused by poor judgment or negligence.

To a prospective employer, the manner in which you conducted yourself in your past employment is a reflection of the quality of employee you will be. Your previous employers will be contacted and many will feel compelled to provide a full review of your job performance.

DON'T BURN YOUR BRIDGES.

There are some ruthless and corner-cutting employers out there. If you can't work in that environment, quit. Before you do, however, follow the checklist below. If you conduct yourself appropriately it will be difficult for an employer to provide negative information during a background check. If you don't conduct yourself appropriately, you may be haunted by an angry employer years after you leave their employ.

Little mistakes add up. Sometimes we don't think about the long-term consequences of certain events until it is too late. People have been turned down because they didn't take their career seriously. They failed to do what common sense dictated at the time. Do not take chances.

✓BE HONEST. You have seen this advice already in this book. I can't emphasize it enough. An employer will fire you in a heartbeat for lying.

✓Treat your employers and coworkers with respect and dignity. Don't forget, your peers can show up to help you, or haunt you, in the future.

✓Give proper notice when leaving a job. If you leave a company without notice, you can count on a bad reference.

✓Sick time is a privilege, not a benefit, and should be used for the purpose it is intended—sickness. An employer wants to hire a dependable employee who will be available when needed. Abuse of sick time is a growing concern among employers. Think twice before calling in sick when you are not. Not only is it unethical, but the long-term consequences could literally make you sick.

✓Don't take unnecessary risks. You may think it would be fun to buzz your home just once in your life. Is it worth the risk? I think not. Use your very best judgment in assessing situations. Before jumping into a potentially damaging or compromising position think about how your actions will impact your future opportunities.

Remember, poor judgment will follow you throughout your career.

EXAMPLES OF POOR JUDGMENT.

✓An applicant forged recommendation letters from acquaintances. The letters were all typed with the same word processor on the same paper and using the same format. The interviewer called one of the references only to find out that person didn't write the letter. An airline will often call all references to obtain more feedback about an applicant.

✓Pilots have attempted to document their experience in a misleading way so the employers think they have more flight experience than they actually do. For example, naval flight officers or weapons systems officers (these are aircraft crew positions but not pilot positions) may show co-pilot time on an application when they really weren't officially qualified to fly an aircraft.

✓Applicants have tried to cover up problems at previous employment by using personal references, or another part-time job, to cover that time frame. The industry is too small. Even if it doesn't show up in your logbooks or in the background check the chances are good the situation will eventually surface.

POOR JUDGMENT FOLLOWS YOU.

✓At age 21 an applicant was violated by the FAA for flying low over a sorority house. Although, at age 30, he had maintained a stellar record since that incident, the incident was still a topic of discussion during his major airline interview.

✓One applicant had two parking tickets—for parking in the same spot! What does that say about his ability to follow the rules? His airline interview was full of questions concerning personal judgment.

ASK THE RIGHT PEOPLE.

✓One of the biggest mistakes you can make is acting on advice given by well-meaning, but unknowledgeable, friends. Only follow advice from the right people:

Ask an airline's medical department about the company's medical requirements, not your friend who just went through the medical exam.

Don't ask an airline pilot who has never interviewed pilot applicants how to act during an airline interview—ask someone who does pilot hiring or a respected pilot interview consultant.

10
NETWORKING

In this highly competitive industry, you cannot sit at home and wait for the jobs to come to you. It just doesn't work that way. You have to take advantage of every resource available to make sure your resume gets seen.

There are plenty of aviation organizations and publications available to guide you in the right direction. Refer to the Appendix 1 for addresses of many of the following resources. See Appendix 5 for Networking Logs.

YOUTH ORGANIZATIONS.

If you are in junior high, high school, or college, consider contacting the following youth organizations. Addresses for these organizations are listed in the back of the book.

Aviation Exploring. This is a division of the Boy Scouts of America. As an Aviation Explorer, individuals between ages 14 and 21 years participate in educational and recreational activities having to do with flying. There are opportunities for individuals of both sexes to learn important skills like team building, public speaking, and interviewing. You might also learn to preflight an aircraft or take pilot training ground school. This has been the first step to a career as a professional pilot for many young men and women.

Civil Air Patrol. The official auxiliary of the U.S. Air Force is a nonprofit, aviation-oriented, volunteer civilian organization. To join, a new cadet must have completed the sixth grade or be 13 years old. The maximum age for entering the Cadet Program is 18. Once a member, a cadet may remain until age 21.

EAA youth and education programs. Sponsored by the Experimental Aircraft Association. They provide many resources and projects available to youth of various ages to generate interest in aviation. Their pilot mentoring program is one of the largest in the industry.

CAREER AND EMPLOYMENT NETWORKING.

Attend aviation conventions and trade shows. This is an excellent way to learn more about requirements of a pilot career and find out which companies are growing. An added bonus of these conventions and trade shows is that you will get to know other pilots.

The following organizations offer career and employment networking to anyone who is interested in aviation:

National Business Aircraft Association
(NBAA). A nonprofit organization formed to promote the aviation interests of corporations in the United States.

Regional Airline Association (RAA).
Nonprofit organization formed to promote the interests of regional and commuter airlines.

Federal Aviation Administration (FAA).
Your local FAA office has brochures that provide loads of information on careers as a pilot. They even have volunteers trained as Aviation Education Counselors located all over the country. Aviation Education Counselors work with schools, individuals, and organizations to provide resources and counseling to assist aspiring pilots. These counselors are members of the industry, including pilots, aviation managers and FAA employees.

ON-LINE NETWORKING.

You can go on-line using your personal computer and network to your heart's content. By accessing the Internet, World Wide Web (web), or on-line services like CompuServe, AOL, and Prodigy you can link up with virtually limitless resources.

By searching on key words, like "aviation," you can access talk forums, placement services, aviation magazines, or aviation industry news.

FOCUSED ORGANIZATIONS.

Organizations like the Women in Aviation International, the Ninety Nines (another world-wide female pilot organization), and the Organization of Black Airline Pilots were developed to help sponsor or support and encourage those pilots who are attempting to meet the challenge of reaching the dream of becoming pilots. These organizations are also accessible on CompuServe and the web.

AVIATION PUBLICATIONS AND SEMINARS.

Check out your newsstand and you will find several magazines on flying. The articles are great and the classified ads have a wealth of information on career opportunities, flight schools, and other services for pilots.

There are companies that provide seminars on airline hiring. The airlines typically send representatives to these seminars to inform the participants about current and future hiring needs. This is a great way to meet the airline people and find out, first hand, what is happening in the aviation industry. Most aviation publications advertise seminar opportunities. As with any decision where money is concerned, make sure you ask for references from people who have attended the seminars.

YOUR FRIENDS.

Once you enter this field, you will find that you seemingly cannot go anywhere without bumping into another pilot. Don't be shy. Ask how they got their jobs. Get your friends to recommend you to their employers or other friends who are flying. The pilot pipeline is one of the best sources of information and feedback.

A WORD OF WARNING.

Although you must be assertive in your networking opportunities—be careful not to be a "networking pest."

Do not take advantage of people who offer suggestions and introductions to others in the industry. Do not call them every week just to chitchat. Do not show up at their office unannounced.

Thank these people for their assistance. If you would like to keep them aware of your progress, do it by writing short notes that you send to them through the mail.

Notes

11

PREPARING FOR THE JOB HUNT

CHOOSING AN EMPLOYER.

M ake sure you choose a company and a job that will help you to improve your flying experience. You always want to be moving forward towards achieving your final goals.

Unless you are using the new job to build hours, check on the company before making a commitment. Many new flight companies don't last long in this competitive industry. Try to get the answers to some basic questions: How long have they been in business? Is their financial status stable? What are the working conditions? Do the pilots working there enjoy flying for the company?

Talk to pilots and other employees to learn as much as possible about the company.

RESEARCH.

Before you send out resumes, analyze the job market and pay special attention to those companies for which you are qualified. Although you should apply everywhere you meet the basic qualifications, your chances may be greater at companies where you have some experience in the type of aircraft they fly. Contact these companies first.

Call the company's public relations department. They will often send an annual report or other information on the company.

Visit the library and review articles about the company and company officers.

Read national aviation related literature or publications. Many of these are now found on-line.

THE RESUME.

There are many reliable references available on how to develop a good pilot resume. The main thing to remember is that you are applying for a position as a pilot, so you must sell your pilot qualifications. The format is not as important as the content, but the format should reflect your professionalism.

Limit your resume to one page. Don't try to be different. Don't use cute pictures or quaint fonts. Supply the facts in a logical and professional way. Use a standard font applying bold print to set off headings or key items.

Some companies use a system that scans resumes. Using unusual fonts such as a script or a handwriting font, may interfere with the ability to pick up information. Keep it simple. In these cases, you want to use as many key aviation terms as possible (i.e., type of aircraft flown, licenses held, type ratings held).

Some airlines or companies use the resume as their only source of information. Therefore you must be sure that you include all aviation related experience. .

Assemble your information using these sections:

✓Your Name, Address, and Phone Numbers.

Include your current and, if applicable, permanent residence. Since you won't be home sitting by the phone all day, be sure to have an alternate phone number (friend or relative) and an answering machine on your main phone number. Don't miss an opportunity because a company can't get hold of you.

Don't put any silly or provocative messages on your answering machine. This is the fastest way to turn off an employer.

✓Your Social Security Number.

Many companies retrieve information and track applicants by their social security number. Place it in a very visible location on your resume. For some people, revealing their social security number is a matter of concern. Be careful how you address this issue with an employer and be prepared to offer an alternative if you don't want to release the number.

✓Objective.

Personalize each resume when you are targeting a specific company. However, you should always have a generic resume available in the event you run into an unexpected opportunity.

✓ Flight Certificates and Ratings.

List all flight related certificates and ratings (include airplanes in which you have type ratings) in reverse chronological order of receipt. If an expiration date applies, include it. If you are scheduled to complete a new certificate or rating, include that information in your cover letter. Be sure to update your resume after you successfully complete the requirements for that certificate or rating. Do not list a certification or rating on your resume until you actually receive it.

✓ Aircraft Flown.

If you have room, list the aircraft you have flown under a separate heading. If not, include them in the job description section.

If you have a great deal of experience and have flown many types of aircraft, you may leave off some of the single-engine aircraft you first flew when beginning your training.

✓ Flying Hours. Break down your hours into the following categories as applicable.

Total time	Single-engine	Flight engineer
Pilot-in-command	Multi-engine	Cross-country
Piston	Helicopter	Night
Turboprop	Instructor	Simulator
Turbojet	Student	Instrument

✓Work Experience.

List all your jobs in chronological order starting with the most recent and working backwards to high school or college. Include:

Month and year of employment.
Company name and address.
Brief description of your responsibilities.

Always include type of aircraft and positions flown, e.g., Captain, Jetstream or Aircraft Commander, KC-135. If you held several positions such as Chief Pilot or Check Airman at a company, list them in chronological order starting with the most recent. Include special assignments.

✓Education.

List all colleges attended in reverse chronological order. If you did not attend college or you haven't accumulated much working experience, include high school. Include:

Dates attended.
Name of college or university.
Major.
Type of degree. If you did not complete your degree list how many credits you have.
GPA. This is optional. If you had great grades, include it.

✓Training/Accomplishments/Job Related Organizations.

Use this section to promote yourself. A company is often interested in special achievements, especially if they are job related. If you participated in flying competitions in college or won an employee of the year award, this is an area to brag a little.

✓ Personal Information.

      Use this section to tell a little about yourself. Many resume services advise you to include date of birth, height, weight, vision and marital status. I recommend you stay away from those items. Companies typically do not use this information in making selection decisions.

Always keep your resume updated.

      You never know who you might meet that knows of an opening for your perfect job. If you meet someone, on the job, who is interested in seeing your resume, send it to them in the mail. Your present employer may not appreciate you looking for another job on company time.

      A sample resume follows.

SAMPLE RESUME.

Ace McFly

Social Security # 123-45-6789

Present Address:	Permanent Address:
100 Airport Rd.	P. O. Box 222
Fly City, CA 12345	Craft, PA 98765
Phone: 555-123-1234	Phone: 555-321-6541
FAX: 555-123-1235	E-mail: 72137,012

OBJECTIVE

Career Pilot Position with XYZ Airline

FLIGHT RATINGS

Airline Transport Pilot, Airplane Multi-engine Land
Flight Instructor, Airplane Single-engine and Instrument
Type Ratings in B-727, A-310, A-320, LR-35
First Class Medical Certificate, no restrictions

AIRCRAFT FLOWN: B-727, A-310, A-320, LR-35

FLIGHT HOURS

TOTAL 7,847

Pilot-in-command	6,632	Multi-engine	4,897	Instrument	1,500
Turbine	4,897	Single-engine	2,950	Simulator	500
Flight Engineer	300	Instructor	2,200	Night	1,435

WORK EXPERIENCE

JOYFUL TRAINING CENTER: Oct. 1997–Present
4333 Ave. 8, Silicon, NV 98765
Captain/Instructor: A310/A320. International Captain, FAA examiner.
Non-scheduled charters.
NORTHEAST AIR: Aug. 1990–Sep. 1997
123 Happy Rd., Melba, CA, 95555
Captain/Co-pilot: B-727, LR-35. Fly passengers on scheduled air service. Some instructor time.

EDUCATION

BS Degree–Professional Pilot Technology, Alpha Aviation University, Alpha, USA.
Member: NIFA, Alpha Eta Rho.
Enjoy intramural sports (baseball, tennis), running, golf.
Excellent health, nonsmoker.
AVAILABILITY: IMMEDIATE, TWO WEEKS PREFERRED

THE APPLICATION.

How you present information reflects your motivation, determination, and attention to detail. A sloppy or incomplete application is likely to go to the bottom of the pile. Neglecting to respond to an important question or making one simple mistake can be the difference between getting an interview or being passed over.

Keep a copy of **everything** you send to a company. This information is useful to review when you are called in for an interview. You will also want to keep a record of whom you sent the information to as well as the date you sent it. You can use this to know when to send in an update.

The application process varies among companies.

The generic company application. Most large companies have their own standard company application that is used for all employees. However, when hiring pilots they typically have a supplement or a section on the application that relates to pilot qualifications.

The scan sheet. You can expect more and more companies to use some kind of form that will allow them to electronically transfer the data to a computer database. This provides them the capability to search and sort through qualifications of the applicant pool.

How you present information reflects your motivation, determination, and attention to detail.

The resume. Most companies have a specific company application. With some smaller companies you may find they use the resume for their main source of information. As a professional you should have a current resume.

Outsourcing. More and more companies are going to services that maintain a large database of pilot qualification information. The company provides specific criteria then retrieves and reviews only those pilot's applications which meet their specific needs. For example, The Universal Pilot Application Service, Inc. (UPAS) is a database service that matches up qualified pilots with companies that have job openings. By submitting your pilot qualification information to the database service, your records will be available to many companies.

Helpful Hints.

It doesn't matter which method a company uses to select their most competitive pilots, the same rules apply when completing the applications or resumes. Here are some helpful hints that might keep you from missing out on that competitive interview. Keep in mind, you still have to meet the "minimum and competitive" qualifications.

✓Read and follow the directions carefully. It is amazing how many people don't. For example, in the work history section you might be asked to start with your current job and work backward. Don't start with your first job and work forward. Pay attention to detail.

✓Unless the company specifically asks you to hand-print the information, **type it**. Most applications are either scanned or manually entered into a database. Even nice handwriting can be hard to read. You will expedite the processing of your application if you type the information. If you are asked to print it, they are looking for your printing skills. Do not have someone else print the application for you.

✓ Fill in all the blanks that apply to you. If you accidentally leave one section blank that is a specific criterion for a search, your name won't show up.

Unfortunately it is not uncommon for applicants to neglect to provide a particular piece of information. In my business, we periodically get calls from applicants who are upset that their data was not released to a company when they think they meet the minimum criteria. When we check the files, we find that they did not fill in the required fields.

✓ A sloppy or incomplete application is also a reflection of your motivation, work habits, and attention to detail, all important factors for flying jobs.

✓ Don't guess. Take the time to do the research and provide accurate and up-to-date information.

✓ Many companies ask you to fill out a flight grid that breaks down your flying time. It seems that each company has a different format and they can be confusing. The interviewers will check this information against your logbooks. As long as they can

READ THE DIRECTIONS!

match up the numbers, you will be all right. Don't guess on the numbers. I can't believe how many pilots round off their hours to the thousands, i.e., 2000 total time, 1000 hours multi-engine time. Come on—we know that pilots don't fly exactly "000" the time! If you have 2043 hours that is what you list, not 2050! Your logbooks and the flight time grid must match.

✓If you have someone else type your application, proofread it carefully. I remember many cases where "no" was marked when "yes" applied. Example: are you willing to relocate? If you mark "no" by mistake, you just lost an opportunity. Pay attention to detail. Make sure the person typing your application is as professional as you are.

✓Remember the company to which you are applying. Sometimes you fill out so many applications you lose track. Pilots have actually filled out the section on "why I want to work for your company" and put in the wrong company name.

✓Include as much information about your experience as space will allow. Use the same principles discussed in the section on the resume.

✓Include a resume and cover letter. Some applications don't give you an opportunity to sell yourself as much as you may wish. So, unless a company specifically tells you not to include a resume or cover letter, send both with your application.

✓What if you don't meet the minimum requirements? Follow the company directions. If they specify that you must meet minimums to apply, wait. However, if they do not set specific criteria, go ahead and apply. You never know when the criteria will change. Be prepared, however, for a response back saying you don't meet the minimums or, you may not hear back at all. On the other hand, if the company maintains a file on you, this type of interest shows a motivation to work for the company.

✓And finally, DON'T USE MAGIC MARKER! You laugh. It has been done. Not only is it inappropriate, but the ink bleeds and you can't read a thing.

THE COVER LETTER.

Is it really necessary to include a cover letter? Unless the company states otherwise, I recommend including a cover letter. It gives you an opportunity to add some extra information about yourself that you might not be able to include in a generic application.

When writing a cover letter keep the following in mind:

✓Find out who is in charge of hiring and direct the letter to that person. This shows you did some research. If you send it to the wrong person you may slow down the process and give the impression that you lack respect for those responsible for the hiring process.

✓Get the name right. I used to get letters addressed to Mr. Tarver or Ms. Traver or Mrs. Carver. Although I especially liked it when they addressed me as Capt. Tarver!

✓Remember which company you are writing. Those fancy word processors can be dangerous if you are not careful. If you are using a form letter to send out lots of resumes and applications and use the airline's name in the copy, be careful to make sure you make all the appropriate changes. It can be embarrassing if you don't. I frequently received letters from applicants addressed to me that had the name of another airline mentioned in the copy.

✓Keep it to one page.

✓Don't forget to add information that cannot be found on a resume or application such as an upcoming certificate.

✓Don't send an application to UPS via FedEx or vice-versa. Big mistake!

UPDATES.

How often should you update? That depends on the company. Some companies want you to wait six months, others one year. Follow their directions. If you are using a database service or employment agency, update as much as you want.

Most companies will not accept telephone updates. They want the information in writing.

Be sure you make it clear to the company why and what you are updating. If you simply send a new resume, they may not take the time to figure it out. Some basic principles to follow are:

✓Definitely update as soon as the company allows if:

   –you change your address or phone number,

   –you change jobs,

   –you complete a college degree,

   –you make a significant change in your qualifications like upgrading to captain, obtaining a type rating, or gaining a certificate that is a requirement for a particular company.

✓If you are a low-time pilot, any new hours are important.

✓You already have plenty of hours, no new job, haven't moved. What then? Update when the company allows to express your continued interest in the company. If they do not specify an update window, update at least every six months.

Notes

12

THE PILOT RECRUITMENT PROCESS

HOW HIRING CRITERIA ARE ESTABLISHED.

You may think you are as qualified as the next person to land that job but your friend, who has a little less flight time than you, gets an interview and you don't. How do companies determine just who does get called in?

Hiring requirements are typically set by experienced flight management pilots. Most have been performing the job for years and have an established track record of success.

Basic requirements can usually be broken down into two categories:

Specific aircraft experience. Some companies are only interested in specific aircraft experience. They want to hire pilots who are already rated in or have experience in the specific aircraft they fly. Training costs are high and if they can find pilots with current experience, they will.

Your total package. The majority of companies, especially the major and regional airlines, look at your entire background, which includes education, hours, job experience, and potential. After a comprehensive selection process the company will train pilots in their own aircraft to their own specifications.

Companies will publish minimum requirements for the job, but actually set the highest qualifications possible according to the current available applicant pool. In other words, you might meet the basic requirements but can still be a long way from being competitive because many other applicants have a great deal more flight time and experience.

SELECTION PROCESS.

The process can be a simple one person interview with the chief pilot or owner of a company or it can be a sophisticated series of interviews and tests.

Although most companies use internal staff to recruit and select prospective candidates, many are turning to professional recruiters or flight training academies to recommend qualified candidates.

When applying to a major airline for a job, one that does regular hiring, you can expect most of the following procedures to occur:

✓ An interview with someone from the human resources or personnel department. These are experienced recruiters who typically are not pilots. They have been trained to evaluate everything with the exception of actual technical flying experience. They will evaluate your career path and your adaptability as an employee. Don't underestimate their importance. They can eliminate you from the process.

✓ An interview with one or more pilots, typically from flight management. They focus on your flying qualifications and other areas that relate to your performance as a crewmember. They will expect you to be an expert on the aircraft you are currently flying and answer reasonable questions about others you have flown.

✓ Testing. Although the results of these tests may result in instant disqualification, they are usually evaluated as a part of the entire package. You may expect to see some or all of the following kinds of testing:

Aptitude. A test to evaluate your overall intelligence. Good scores on these kinds of tests usually indicate a higher success rate in training.

Psychological. Most of these tests are custom made for the employers. They are developed to find pilots who are most like the company's own successful employees.

Cognitive. Tests that evaluate your ability to multi-task and work in complex environments.

Technical. These tests focus on your flying knowledge. They evaluate your generic knowledge about basic flying principles and FAA rules, etc.

✓Simulator Evaluation. This is a test of your basic flying and instrument skills. You might be put in a large full-motion simulator or a non-motion computer scored trainer. In some companies, you can be eliminated immediately if you fail, or do poorly, during the simulator check.

FINAL SELECTION.

Once all the information from the interviews and testing is compiled, the information is reviewed once more by the person or people who will make the final hiring decision.   Selection can be made at the interview stage by those conducting the interview or it may be made by a board of review composed of management people who make the decision based upon the data compiled throughout the interview process.

If a positive selection decision is made, a candidate is extended a conditional job offer contingent upon passing a company medical and a background check.   Some companies' medical examination is satisfied if you can present a current FAA First Class Medical Certificate, others require a more comprehensive company physical. The FAA requires a background investigation be completed on each new-hire. That includes obtaining verification of employment at least five years back and sometimes, depending upon the circumstances, ten years.

Your FAA records will be checked. New legislation, effective February 6, 1997, requires companies to retrieve training records on new-hire pilots for the last five years. They also require retrieval of driving records and verification that you maintain the appropriate federally mandated medical certificates. You must be prepared to provide complete employment information or you may hold up the process or possibly miss an opportunity altogether. A company may also conduct:

- ✈ Criminal background check
- ✈ Education verification
- ✈ Credit check

Notes

13

THE INTERVIEW

Whether this is your first interview for a single-engine flight instructor position or an interview with a major airline, treat it with the same vigor and determination.

In most cases, you only get one chance. Don't blow it. Be prepared. Each job is one step closer to meeting your ultimate dream goal. If you prepare correctly the odds are you will reach your goals more quickly.

PREPARATION.

First, do everything I have been telling you throughout the book! Then have everything in order for your interview.

I strongly recommend taking advantage of an interview skills preparation course. It will help lower your stress level during your interview and will bring out the best in you. This is especially helpful if you have a black mark on your record. A good interview consultant can teach you how to turn a negative into a less objectionable situation.

A few points to remember:

✓ Don't be tired or sluggish for the interview. Get plenty of sleep and eat right.

✓ Stay away from pungent foods and alcohol for at least three days prior to your interview.

✓ Have all of your documents available for review. Remember, pay attention to detail. Refresh your memory about the information you sent to the company. This is where the copies of all documentation you have saved will come in handy.

✓ Make sure you can answer any technical questions about the aircraft you are flying or have flown.

✓ Make reliable travel plans to get to the interview and be sure to arrive early.

APPEARANCE.

This is my absolutely favorite thing to talk about. Appearance is not only what you wear, but how you present yourself. First impressions can impact the entire

Make sure the interviewer is focused on your qualifications, not what you are wearing.

interview. Interviewers will observe your dress, grooming, and hygiene.

I remember an interview when a female applicant wore a very provocative outfit. When it was over, I guarantee, the interviewers did not remember one thing about her qualifications. She was the talk of the office for two days, but not for the right reasons.

✓ Don't wear excessive jewelry.

✓ Wear your hair in a conservative cut or style.

✓ Don't wear perfume or cologne. Some people are allergic. The only thing the interviewer will be thinking about is when the applicant will leave so he or she can breath again.

✓ Do bathe and use deodorant. Do brush your hair and teeth. Again, you think I am kidding. Not so! I have interviewed people who had such bad body odor that it was difficult to concentrate.

✓ Don't wear funny suits or ties. You don't have to wear the typical blue suit, white shirt, red tie, and black shoes. Just be sure your suit is in style and conservative.

✓ Keep the attention where it belongs, on the interview. Allow the interviewer to remember you, your strengths, your qualifications and qualities—not what you were wearing.

Many of you are probably thinking you should be able to express your individuality. I wouldn't recommend it. Remember, you are going to have to wear a uniform and conform to some very strict standards when you represent the company in public. By wearing inappropriate attire at an interview, you are saying, "I won't conform." You are sending a negative message about your attitude to the interviewer.

BEHAVIOR.

Your behavior will be observed from the minute you answer the telephone when you are scheduling your interview until the final recommendation is delivered.

Here are some tips:

✓ Treat each person you deal with, whether it is the ticket agent at the gate, the secretary, or the interviewer, with respect and dignity. Bad gossip always travels faster than good.

✓ Don't ever assume you are safe from being seen or heard. I remember a time when an applicant just completed the interview and was in the van going to the airport. He proceeded to complain about the recruiter and the interview process. Guess who was sitting in the back seat of the van?

✓ Don't be condescending. Unless you are asked, assume the company representatives know what they are doing.

✓ If you have a recorder on your phone, make sure the messages are professional. A provocative or silly message will turn off an airline representative quickly.

✓ An interview is not the time to crack jokes. Humor might add to an interview at certain times, but it can be risky.

✓ If you know that the person interviewing you is not a pilot, please don't ask him or her what their qualifications are to interview you.

Inappropriate Behavior.

The following are some examples of inappropriate behavior from my experience of interviewing. Some were displayed onsite, some offsite.

- When greeting an applicant in my office for an interview, I held out my hand to shake his and he handed me his briefcase and sat in my chair. He was condescending the whole time.
- While waiting in the office for his interview, one fellow took off his jacket, loosened his tie, put his feet up on the chair nearby, and asked the secretary if she would share her lunch.
- An applicant flying first class on his pass home got drunk and disorderly.
- An applicant brought a soft drink into the interview.
- Gum chewing occurs on a regular basis and is very distracting.

FOLLOW UP.

Let the employer know that you want the job. Write a nice thank you note expressing your thanks for the interview and your sincere interest in working for the company.

Frankly, although a thank you note has become expected with major airlines, it has more value in a small company.

I have also found that those pilots just starting out don't seem to think of thank you notes. However, it made the difference for many of the applicants I know who might have been overlooked had they not followed up with a note.

REJECTION.

What happens if you get turned down? Keep trying. The pool of pilots changes. You might be unsuccessful today, but could be on the top of the list in a couple of months. Pilots are often reconsidered. Find out how long you must wait to reapply and send in an update. Just because you were turned down by XYZ company does not mean you will get turned down by ABC. You must persevere and keep trying.

Most companies will not tell you why you were not selected. Don't push the issue. You will only get a nice letter saying that they won't tell you why you were not selected.

This all goes back to my first comment to you in the book. Don't go into this career if you have your sights set only on a particular company.

I have been turned down for jobs for which I felt highly qualified. I chalked it up to experience and kept on going.

Evaluate how you did in the interview. Debrief your interview experience with someone who understands the pilot interviewing process: a former interviewer or a pilot interview consultant.

14

CLOSING

Now you have a broad overview of how you can proceed with your plans to become a professional pilot. Take advantage of the many resources available to you to fine tune that task.

Looking into the future, predictions for job opportunities are excellent well into the first ten years of 2000. Airlines are continuing to expand and many major companies will lose thousands of crewmembers due to retirement and attrition.

I cannot emphasize enough the importance of following the advice in this book. Do not take these career requirements lightly.

✓Becoming a professional pilot requires time, money, dedication and sacrifice.

✓Make a plan, set a timeline. Do not get sidetracked.

✓ Study hard and ace your aviation related courses and tests. Do well in your non-aviation courses.

✓ Keep accurate and up-to-date records.

✓ Take advantage of all the resources available to you.

✓ Do your research on potential employers.

✓ Don't burn your bridges.

✓ Take care of your health.

✓ Always be prepared. Never wait until the last minute to gather documentation for potential interviews. Don't take shortcuts.

✓ Be honest...Be honest...Be honest.

I hope your flight plan to the flight deck is successful as well as fulfilling and rewarding. Good luck!

APPENDIX 1

Resources

APPENDIX 1

Resources

NETWORKING

Alpha Eta Rho
Executive Director
1615 Gamble Lane
Escondido, CA 92029-4326
760-737-8308
760-746-2863 Fax
Kenbac@aol.com
http://www.unomaha.edu/~ahp/ahp.html
National Aviation Fraternity.
Provides scholarships to members.
Provides networking opportunities in aviation for members.

Aviation Exploring (Division of Boy Scouts of America)
National Director
P. O. Box 152079
Irving, TX 75015-2079

972-580-2000
http://www.bsa.scouting.org/programs/02-504.htm
Non-profit organization for youth between 14 and 21 that
focuses on careers in aviation.
National Conferences dedicated to aviation.

Civil Air Patrol (CAP)
105 South Hansell Street, Building 714
Maxwell Air Force Base
Alabama 36112-6332
800-FLY-2338
Cadet Program. Official Auxiliary of U.S. Air Force.
Http://www.cap.af.mil
Stimulates public interest in aerospace issues.

Experimental Aircraft Association (EAA)
PO Box 3086
Oshkosh, WI 54903-3086
920-426-4800
http://www.eaa.org
EAA Young Eagles Flights.
EAA Jr. Air Academy - Ages 12-14.
EAA Air Academy - Ages 15-17.
EAA Project Schoolflight.
EAA Scholarship Program.
Sports Aviation Magazine.

National Intercollegiate Flying Association (NIFA)
PO Box 3204
Delta State University
Cleveland, MS 38733
601-846-4208
Conducts regional and national flying competitions among
two and four-year aviation colleges and universities.

Ninety-Nines
Will Rogers Airport
PO Box 965
7100 Terminal Drive
Oklahoma City, OK 73159-0965
405-685-7969
http://www.ninety-nines.org
Provides mentoring programs for women in aviation.
Publishes *International Women Pilots* magazine.

Organization of Black Airline Pilots (OBAP)
P. O. Box 50666
Phoenix, AZ 85076-0666
800-JET-OBAP
http://www.obap.org
CompuServe (GO OBAP)
Provides Pilot Development Program, scholarships, pilot
mentoring.

Women in Aviation
Morningstar Airport
3647 S.R. 503 S.
W. Alexandria, OH 45381
937-839-4647
http://www.wiai.org
Provides year-round resources to assist women in aviation
and to encourage young women to consider aviation as a
career through educational outreach programs.
Publishes *Aviation for Women.*

SERVICES

Cage Consulting, Inc.
13275 E. Fremont Place, Suite 315
Englewood, Colorado 80112
Toll Free:1-888-899-CAGE (2243)
Local: 1-303-799-1991 and FAX 1-303-799-1998
www.cageconsulting.com

Offers Interview Preparation Services in addition to many
excellent Pilot Interview Study Guides including:
*Checklist for success: A Pilot's Guide to the Successful
Airline* by Cheryl A. Cage *CHECKLIST Interactive CD:
An Interview Simulator* by Cheryl A. Cage *Airline Pilot
Technical Interviews: A Study Guide* by Ronald McElroy
*Flight Plan to the Flight Deck: Strategies for a Pilot
Career* by Judy A. Tarver just to name a few.

Universal Pilot Application Service, Inc. (UPAS)
751 Miller Drive, Suite D2
Leesburg, VA 20175
800-PILOT AP
http://www.upas.com
Pilots enter their qualifications in a master database that is
used by companies looking for candidates.

ASSOCIATIONS

Aircraft Owners and Pilots Assoc. (AOPA)
421 Aviation Way
Frederick, MD 21701
800-USA-AOPA
http://www.aopa.org
CompuServe (Go AOPA)
AOPA Project Pilot
AOPA Project Pilot Mentor Program
AOPA Project Pilot Instructor Program
AOPA Pilot Magazine.
AOPA Airport Directory - The Pilot and FBO Planning Guide.
Flight Training magazine.

Air Line Pilots Association
Pilot Information Program (PIP)
535 Herndon Parkway
Herndon, VA 22070
703-481-4440
http://www.alpa.org
Publishes *Airline Pilot Career Information*

Federal Aviation Administration (FAA)
http://www.faa.gov (FAA home page)
Provides a variety of resources on aviation.
Publishes *Aviation Careers -The Sky's the Limit.*

General Aviation Manufacturers Association (GAMA)

1400 K Street, NW, Suite 801
Washington, DC 20005
202-393-1500
http://www.generalaviation.org
Learn to Fly - A brochure that answers preliminary questions about flying.
Your Career in General Aviation - a book describing job opportunities.

National Air Transportation Association (NATA)

4226 King Street
Alexandria, VA 22302
800-808-NATA
NATA is a public policy group representing the business interest of aviation companies engaging in flight training. In coordination with *Flight Training* magazine, they publish a pamphlet titled ìChoosing a Flight Schoolî.

National Association of Flight Instructors (NAFI)

PO Box 3086
Oshkosh, WI 54903
920-426-6801   membership phone 800-843-3612
http://www.nafinet.org
Provides comprehensive resources for flight instructors.
Only source of flight instructor liability insurance.
Flight instructor Hall of Fame.

NAFI Mentor magazine

National Business Aircraft Association (NBAA)
1200 Eighteenth Street., NW
Washington, DC 20036-2506
202-783-9000
http://www.nbaa.org
NBAA is an association dedicated to increasing the safety
and efficiency and acceptance of business aviation. They
represent over 4,100 companies.
Publish a brochure titled *Careers in Business Aviation.*

Regional Airline Association (RAA)
1200 19th St. NW, Suite 300
Washington, DC 20036-2422
202-857-1170
http://www.raa.org
e-mail: raa@dc.sba.com/
Represents the interests of regional members on regulatory
and legislative issues

University Aviation Association (UAA)
3410 Skyway Drive
Auburn, AL 36830
334-844-2434
www.vaa.auburn.edu
Publishes the *Collegiate Aviation Directory*
Scholarship listings.

PUBLICATIONS

Air Line Pilot Magazine
535 Herndon Parkway
Herndon, VA 20170
800-448-5313
Http://www.alpa.org

Airline Pilot Careers Magazine
3800 Camp Creek Pkwy #18-100
Atlanta, GA 30331
http://www.airapps.com

Aviation Week and Space Technology
1221 Avenue of the Americas
New York, NY 10020
800-525-5003
http://www.awgnet.com
CompuServe: GO AWG

Business and Commercial Aviation Magazine
4 International Drive
Rye Brook, New York 10573
Subscription 800-257-9402
Fax 914-939-1184
http:www.awgnet.com

Flight Training Magazine
201 Main Street
Parkville, MO 64152
816-741-5151
Published by AOPA

Flying Magazine
500 West Putnam Ave.
Greenwich, CT 06830
203-622-2700   subscriptions 800-678-0797
203-622-2725 Fax
e-mail: hfmflying@aol.com

International Women Pilots Magazine
Official Publication of the Ninety-Nines, Inc.
4300 Amelia Earhart Rd.
Oklahoma City, OK 73159
405-685-7969
http://www.ninety-nines.org

Professional Pilot Magazine
3014 Colvin Street
Alexandria, VA 22314
703-370-0606
703-370-7082 Fax
http://flightdata.com/propilot
Pro Pilot: Published for 31 years. Articles on
safety and technique written by active aviation
professionals for pilots in the corporate, regional,
charter and major airline areas of aviation.

Woman Pilot Magazine
P. O. Box 485
Arlington Heights, IL 60006-0485
Subscription Order # 800-300-7343
Other calls 847-797-0170
Fax: 847-797-0161
www.womanpilot.com

Women in Aviation
The Publication
P. O. Box 40
Lake Ann, MI 49650-0040
616-275-6266
Http://www.wiai.org

WHAT EVERY PILOT WONDERS:
Common Career Questions

1) What criteria should I use in choosing a new job?

That will depend upon whether your next step is the majors or a company which will help you build experience towards that goal.

Major Airline
Prior to deregulation, pilots would jump at the first major airline that offered them a job. However if you are looking for a major airline job, you now need to seriously evaluate: company stability, compensation and benefit packages and corporate culture. Research the history of the airline industry. After deregulation many large airlines that had been around for decades went bankrupt. Keep in mind that you have to start over each time you change airlines. Learn about the relationship between the pilot group and the company. If there is a union involved, learn about the company's union background.

I know a pilot who went to work for a company and was there for ten years when he finally decided that the company culture just wasn't one in which he was comfortable. Now, after ten years, he is trying to make a decision as to whether he should start over. Do your homework *now,* at the beginning of your career to avoid finding yourself in this type of situation.

Career building job
If you are choosing a job as part of your career path toward your final goal, then you need to choose a company that will provide you an opportunity to fly often, upgrade quickly, and fly more and more complex equipment.

Look at the integrity of the company by talking to present and former pilot employees. Do not go to work at a company that might be known to try to make pilots push the safety window. Make sure you are always improving your qualifications along the way.

2) What should I do if I discover that my employer's maintenance practices are sloppy?

You must always work on the side of safety. Try to be as diplomatic as possible in maintaining your stance in not violating FAA regulations, or compromising safety… and then start looking for another job.

Be very careful not to burn your bridges. If the employer runs a less-than-stellar operation, then it is likely that he or she will also be vindictive when it comes to referrals.

However, the best advice is to DO YOUR RESEARCH on these types of situations prior to accepting a job!

3) Why do I need to keep logbooks after I receive my ATP?

You should never stop keeping logbooks. Over the years I interviewed many pilots who worked for MAJOR airlines that went defunct. They never thought they would have to show their logbooks again. Unfortunately, that was not the case. In order to complete the interview process, they had to go back (some of them years) and recreate their flight logs. You cannot predict the future. It is easier to keep your logbooks current than have to recreate years of flights.

4) Would an advanced degree (masters and up) help me receive an interview faster?

Getting an advanced degree is a nice thing to do, but it really won't get you an interview any faster.

It certainly shows initiative and might help you after you get the job if you have a desire to get into the management side of the airline.

But, to receive a job offer to be a line pilot, the airlines strongly prefer the undergraduate degree and the rest of the process really focuses on your flying experience.

5) Why are the airlines so worried about my GPA, especially in non-aviation related subjects?

GPA is important because it shows a dedication to overall success. It would indicate that you will be more likely to do well in training.

6) What is considered job-hopping?

In the process of trying to progress through a pilot career, most pilots usually have numerous jobs. Airlines expect that. What the airlines *don't want* to see is someone who changes jobs frequently without good reason. You should only change a job if you will be making a positive change: gaining more flight hours, flying bigger/faster aircraft, upgrading to PIC, etc.

If you decide you really want to move back to your hometown and you go from a great corporate job flying a jet back to a single engine charter job at lower pay, a red flag will go up with the airlines. This type of decision will minimize their interviewer's view of your career motivation.

7) What type of extracurricular activities, in college and even after graduation, should I participate in?

Being involved in aviation related extra curricular activities such as the flying team, Alpha Eta Rho, etc., is certainly a plus. But it is just as important to be involved in other activities which give you the opportunity to interact with your peers and develop your leadership and interpersonal skills: volunteer organizations that organize fund raisers, team sports, etc.

Be careful not to spread yourself too thin when you join these types of organizations. Remember your first priority is to do well in your flight training and college courses. It would be better to join one organization and become very involved with that club, versus joining numerous organizations and not be able to devote enough time to any of them.

8) Why are the airlines so concerned about my driving record?

Your driving record is another area which helps the airline gauge your judgement. Although everyone has probably had one speeding ticket sometime in their life, the airlines are looking for trends especially if it involves a DUI. With the heavy public scrutiny on these issues in the last few years, anyone who continues to exercise poor judgement while driving is a poor risk.

Your driving record is well within your control. There is no excuse for someone who wants to be a professional pilot, which means being responsible for thousands of lives over the course of their career, to have a poor driving record.

9) I have a poor driving record. What can I do to redeem myself in the eyes of the airline interviewer?

The best thing that will help to overcome a poor driving record is TIME. If you have a DUI you may need up to five years of a perfect driving record in order for the interviewer to become comfortable that this is not a trend.

If you are lucky enough to get the interview, understand that the airline is already aware of your driving record. The main focus will then be how you discuss the situation.

I recommend professional interview preparation to help you work through your discussion of the situation in the interview environment. You must accept reponsibility for the situation, tell what you learned, then sit back and answer all of their questions.

10) I have been fired from a job. What can I do to redeem myself in the eyes of an airline interviewer?

There are a couple of issues here to address. Why were you fired? Was it for poor performance, misconduct, or a personality conflict? Everyone's story is different. Be prepared to discuss the issue as unemotionally as possible. DO NOT blame your problems on anyone else. Take responsibility for your actions, discuss the lessons you learned and what you have done since then to overcome the obstacle.

Having good references from people who worked with you since the firing, and having good referrals from pilots who work for the airline, will also help.

11) What is more important: PIC or jet time?

There are lots of variables in this situation.

For example, many new-hire pilots come from regional carriers which indicates the value of the multi-engine PIC experience received in that environment. When the airlines look at PIC time, they are typically reviewing multi-PIC.

If you have a great deal of jet first officer time and you have the opportunity to upgrade to PIC in a turbo prop, it might be a good decision to take the PIC job.

It is important to know what the qualifications are to GET AN INTERVIEW. Good research will help you make that decision.

12) Is it a good idea to get a non-flying job within the major airline I want to work for and try to move into flight operations laterally?

I do not recommend getting a non-flying job with an airline

simply to get your foot in the door. If you are a good employee, that may give you some consideration. However, the airline is always going to be most interested in your flying experience and competitive qualifications.

Doing a non-flying job will, obviously, limit your flying jobs and take you much longer to build up your flying experience. I know people who have taken this route and were turned down by their company as a pilot because they were behind their peers in experience. It was quite a set-back for them.

13) Should I get my training through military or the civilian route?

It is strictly a personal choice. Please refer to the chapters in this book for guidance on that issue.

14) Everyone talks about networking. What exactly does it mean and how do I do it?

Networking simply means becoming involved in your career field. You will be surprised at how many people you will meet, how many jobs you will hear about, simply by making sure you "stay in the loop."

By this I mean, do some research on the various aviation organizations in your area, and join the one that seems to have the most active membership. Volunteer to assist with projects and fund raisers. Go to the various seminars that the FAA sponsors, attend college aviation career fairs.

Also, remember to be helpful to others. If you can share information or personal experience advice with other pilots, don't be stingy. It is true that "what goes around, comes around!"

15) I have had some problems with my intial training, specfically failed my private and my instrument. How can I

make sure this does not happen again? Also, how will these failures impact my long-term goal of working for a major airline?

Review *where* in your checkrides you failed. Was it in the flying part of the check or during the oral? Perhaps you need to over-practice until you can do all the maneuvers in your sleep. Or perhaps you might need to reevaluate your study habits.

Also, are overly nervous the day of the checkride? If so, take some steps that can help you get over "checkride-itis." Visualize yourself taking the checkride and doing well. Sit down and talk with others who have passed the checkride you will be taking and have them walk you through their experiences. Perhaps they have some suggestions that will help you through your checkride.

You must understand that you have a great deal of control over the outcome of your checkrides. If you study, see yourself succeeding, and know that you have taken every step you can to be prepared, you will be much calmer. If you feel prepared it will be easier for you to focus on the task at hand.

As far as the airlines are concerned, once again they are looking for trends in behavior. If you do well on your checkrides from now on, by the time you get to a major airline interview you will have proven yourself and they will not be concerned about your ability to pass checkrides on the first try!

Notes

Notes

Notes

Notes

Notes

Index

A

B

C

D

BOOKS BY CAGE CONSULTING, INC.

Checklist for Success:
A Pilot's Guide to the Successful Airline Interview
by Cheryl A. Cage. A guide to understanding, and preparing for, the competitive professional pilot interviewing process. A must for anyone in the pilot job search.

CHECKLIST Interactive CD:
An Interview Simulator
by Cheryl A. Cage. This CD offers examples of some of the most commonly made interviewing mistakes and the positve scenarios you will experience after working through this invaluable CD! Just like being in the interview!

Airline Pilot Technical Interviews:
A Study Guide
by Ronald D. McElroy. With over 23 years of airline experience, Ron McElroy guides you through mental math, approach plates, weather, AIM, FARs, and cockpit situations to analyze and present. Hone your flying background into a stellar interview experience.

Also published by Cage Consulting, Inc.

Can You Start Monday? A 9-Step Job Search Guide —
Resume to Interview
by Cheryl A. Cage. A step-by-step job search and interviewing guide for the new graduate or anyone returning to the work force.

Welcome Aboard! Your Career as a Flight Attendant
by Becky S. Bock (with Cheryl A. Cage). Learn about the responsibilities of the professional flight attendant. Get all the information you need to become prepared fothe competitive flight attendant interviewing process.

TO LEARN MORE ABOUT CAGE CONSULTING
PRODUCTS AND SERVICES
or
to *order any book on-line* please visit our website at
www.cageconsulting.com
or call us toll free at 1-888-899-CAGE (2243)

RETAILER DISCOUNTS AVAILABLE